Created Just For You By

Greg Horn
Carmen S. Horn
lisa eiler

Perhaps you know that a **COW** says **Mooo!**

And a **ROOSTER** says

Cock-a-Doodle-Doo!

BUT did you know

That a **COWSTER** says

Moodle-
Doodle-
Doo!

Of course you know that a COW says Mooo!

And a **FROG** says

Ribbit!

BUT you may not know

That a **CROG** says

Moobit!

You already know
that a COW says Mooo!

And a KITTY says

Meow!

BUT could you know

That a **COWTTY** says

Meeooooo!

I know you know
that a COW says Mooo!

And a **MONKEY** says

Hoo-Hoo-Hoo-Ha-Ha-Ha!

BUT should you know

That a **MONKOW** says

Moo-hoo-ho-
ha-ha-ha!

By now you know
that a COW says Mooo!

And a DOG says
Roof!

BUT would you know

That a

COG

says

Moof!

I bet you know that a COW says Mooo!

And a CHICKADEE says

Chickadee-dee-dee-dee!

BUT might you know

That a **COWADEE** says

Moowadee-dee-dee!

You have to know that a **COW** says **Mooo!**

And an
OWL says

Who-who!

BUT did you know

That a **COWL** says

Moo-who!

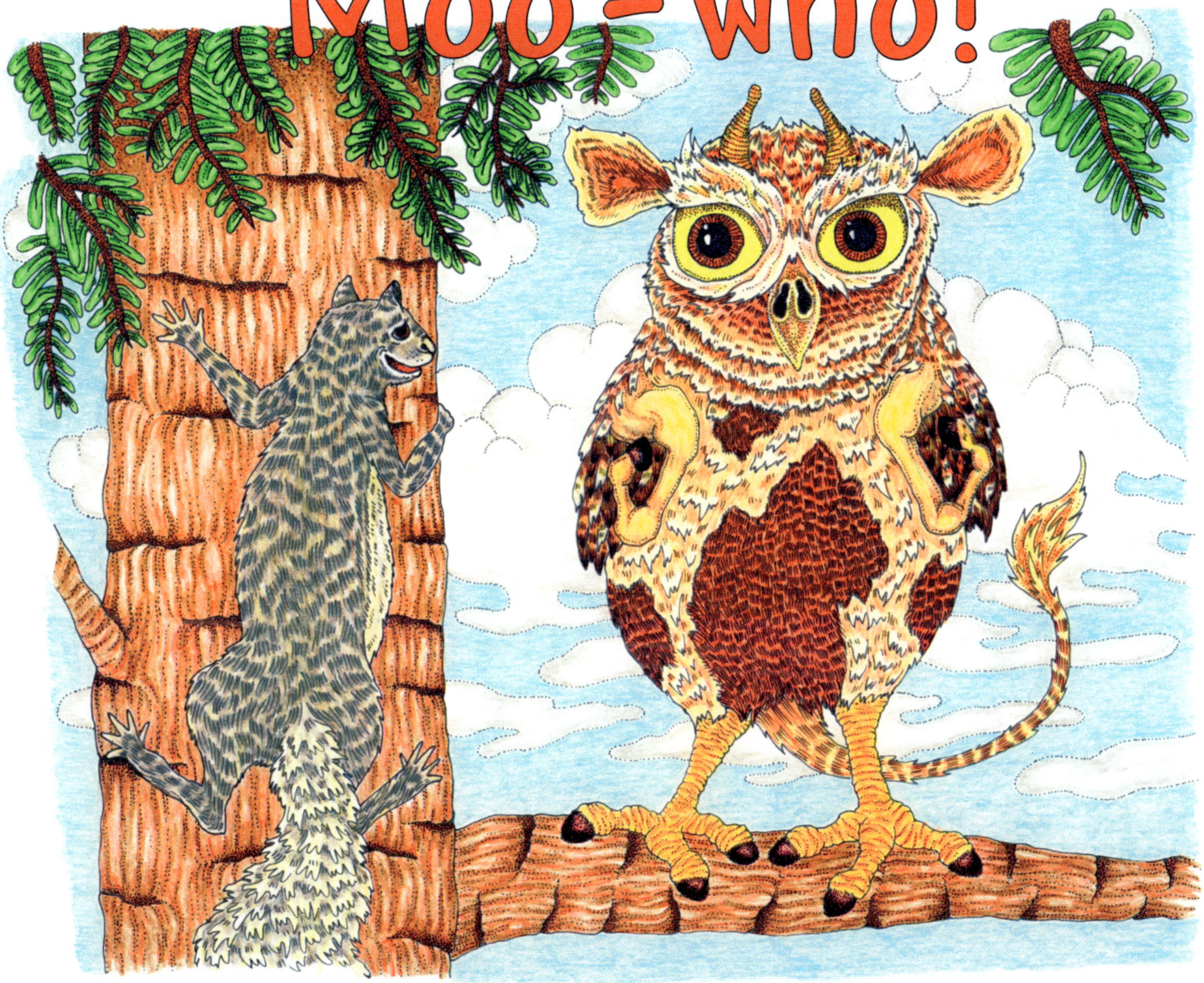

I'm sure you know that a COW says Mooo!

And a **WOLF** says

Aooowww!

BUT could you know

That a **COWOLF** says

Maoowww!

You better know that a **COW** says **Mooo!**

And a
CHICK says

Peep-
Peep!!

BUT should you know

That a **CHICKOW** says

Moop Moop!

THE END

Pheooow!

Pheeeeewww!

About the Authors

Greg Horn (also known as Dadn) lives with his family in Lighthouse Point, Florida. This book was a team project with his very funny daughter Carmen, who at age two loved making up new animals that made sounds no one had ever heard before. After reading countless cow books together, these were refreshing, and this silly book was born.

About the Illustrator

Lisa Eiler is a fine artist and illustrator living in southern New Jersey. She specializes in children's illustration and wildlife art. She is an avid bird watcher and finds the pine barrens of New Jersey to be a paradise that supports her passion for art and nature.

Made in United States
Cleveland, OH
05 November 2026

47649167R00026

Flamingos